Minnesota images/covers/$3_{p}ractice_{t}ests.jpeg 4.3in 4.3in 065 NavyBlue$

AF482328

3 Minnesota MCA Grade 3 Math Practice Tests

Full-Length Test Prep with Detailed Answer Explanations

Dr. A. Nazari

Copyright © 2026 Dr. A. Nazari

Published by View Math Education

ViewMath.com

3 Practice Tests to Get You Started!

Hey there, future math whiz! ⭐

*This book has **3 full practice tests** to help you warm up for the real thing. Think of it like stretching before a big game — these tests will get your brain ready and show you what to expect!*

👍 *Three tests is the **perfect start**!*

👍 *Each one helps you feel **more ready**!*

👍 *You'll be surprised how much you **already know**!*

Sharpen your pencil and let's get warmed up! 🔥

> 66 *Three practice tests is a great way to start. Take your time with each one, and you'll feel more confident every step of the way!* 99

 How to Use This Book

Your quick-start guide to 3 great practice tests!

What's Inside This Book

- **3 Full-Length Practice Tests** — Each one covers all the Grade 3 math topics you need to know!
- **Answer Key with Explanations** — Find out why each answer is correct, not just what the answer is.
- **Reference Pages** — A math symbols chart and multiplication table you can peek at any time.
- **A Test Tracker** — Write down your scores and watch your confidence grow!

A Simple 3-Test Plan

With just 3 tests, here's a great way to use them:

- **Test 1** — **The Warm-Up.** Take this test without a timer. Get comfortable with the question types. Don't worry about your score — just do your best!
- **Test 2** — **The Practice Round.** After reviewing Test 1, try this one with a timer (ask a grown-up!). Focus on the topics that were tricky last time.
- **Test 3** — **The Real Deal.** Treat this like the actual test: quiet room, timed, no peeking at answers. See how much you've improved!

Multiple Choice

Pick the **one best answer** from choices A, B, C, or D. Not sure? Cross out the ones you know are wrong, then pick from what's left. That's a smart move!

Short Answer

Write your answer **and** show your work! Even if your final answer isn't right, showing your steps can earn you credit. Use scratch paper if you need more room.

66 After Each Test 99

Flip to the Answer Key and check your work. For every question you got wrong, **read the explanation carefully**. Then write the tricky topics on your Test Tracker page. If you need extra help, grab our **Grade 3 Math Study Guide!**

⭐ ***Fun fact:*** *Three tests is all it takes to see real improvement! Most kids feel way more confident after just a few rounds of practice.* ⭐

Get Online

Find more at
ViewMath.com/MN-Grade3

ViewMath.com

 Tips for Test Day

Easy tricks to help you feel calm and do your best!

🌙 The Night Before

- ✅ **Sleep early** — *your brain learns while you sleep!*
- ✅ **Pack your supplies** — *pencils, eraser, scratch paper, all ready to go.*
- ✅ **Tell yourself:** *"I've been practicing. I'm going to do great!"*

👉 5 Simple Rules for Every Test

1. **Read the question twice.** *The first time to understand it. The second time to catch details.*
2. **Show your work.** *Write the steps down, even on scratch paper. It helps you think!*
3. **Skip the hard ones.** *Put a small star next to tricky questions and come back later. Answer the easy ones first!*
4. **Never leave a blank.** *For multiple choice, your best guess is better than no answer at all.*
5. **Check your work.** *Finished early? Go back and re-read your answers.*

✅ Smart Moves

- Take a deep breath before you begin
- Underline key words in the question
- Use drawings or number lines to help
- Cross out wrong answers first
- Double-check addition and subtraction

❌ Traps to Avoid

- Rushing and not reading carefully
- Picking the first answer that "looks right"
- Forgetting to carry or borrow numbers
- Skipping a question permanently
- Panicking when you see a tough problem

> ❝ *Remember, the very first practice test is the hardest — not because the questions are harder, but because everything is new! By Test 3, you'll feel like a pro. Trust me!* ❞

 Get Ready to Practice 

Here's everything you need before you start!

Pencils

Sharpened and ready!

Eraser

Everyone makes mistakes!

Scratch Paper

For working things out

A Calm Spot

Somewhere quiet to focus

A Grown-Up

To help set a timer

A Can-Do Attitude

You've totally got this!

✓ Allowed During Tests

- Pencils and erasers
- Blank scratch paper
- The **reference pages** in this book
- A ruler (for measurement questions)

✗ Not Allowed

- Calculators
- Phones, tablets, or computers
- Help from anyone else
- Your study guide (save it for after!)

For Parents & Teachers

- With only 3 tests, **space them at least a week apart.** This gives time to review mistakes before trying the next one.
- Let your child take Test 1 untimed to build familiarity.
- After each test, go through the Answer Key together. Focus on **understanding the "why,"** not just the score.
- If a topic keeps tripping them up, review it in our **Grade 3 Math Study Guide** before the next practice test.
- Celebrate every bit of progress — even getting one more question right is a win!

X¹ Math Reference Sheet X¹

Symbol	Name	What It Means	
$+$	Plus (Add)	Put numbers together.	$3 + 5 = 8$
$-$	Minus (Subtract)	Take away from a number.	$9 - 4 = 5$
$\times$	Times (Multiply)	Add equal groups.	$4 \times 3 = 12$
$\div$	Divide	Split into equal groups.	$12 \div 3 = 4$
$=$	Equals	Both sides are the same.	$2 + 3 = 5$
$>$	Greater Than	The left number is bigger.	$7 > 3$
$<$	Less Than	The left number is smaller.	$2 < 9$
$\frac{1}{2}$	Fraction Bar	Part of a whole.	$\frac{1}{2}$ means 1 out of 2 equal parts

📖 Key Math Words

- **Sum** — the answer when you add
- **Difference** — the answer when you subtract
- **Product** — the answer when you multiply
- **Quotient** — the answer when you divide
- **Factor** — a number you multiply
- **Array** — objects in rows and columns
- **Fraction** — a part of a whole
- **Numerator** — the top number in a fraction
- **Denominator** — the bottom number
- **Equation** — a math sentence with =
- **Estimate** — a smart guess, close to the real answer
- **Perimeter** — the distance around a shape
- **Area** — the space inside a shape
- **Rounding** — making a number simpler by going to the nearest ten or hundred

🔍 Word Problem Clue Words

- **Add** (+): in all, total, altogether, combined, sum, both, more
- **Subtract** (−): how many more, how many left, fewer, difference, remain
- **Multiply** (×): each, every, groups of, times, rows of, per
- **Divide** (÷): share equally, split, each group, how many groups, per

Get Online

Find more at
ViewMath.com/MN-Grade3

▦ Multiplication Table ▦

You may use this table during your practice tests!

×	1	2	3	4	5	6	7	8	9	10	11
1	1	2	3	4	5	6	7	8	9	10	11
2	2	4	6	8	10	12	14	16	18	20	22
3	3	6	9	12	15	18	21	24	27	30	33
4	4	8	12	16	20	24	28	32	36	40	44
5	5	10	15	20	25	30	35	40	45	50	55
6	6	12	18	24	30	36	42	48	54	60	66
7	7	14	21	28	35	42	49	56	63	70	77
8	8	16	24	32	40	48	56	64	72	80	88
9	9	18	27	36	45	54	63	72	81	90	99
10	10	20	30	40	50	60	70	80	90	100	110
11	11	22	33	44	55	66	77	88	99	110	121

💡 How to Use This Table

To find **4 × 7**:

1. Find **4** in the left column (blue).
2. Find **7** in the top row (blue).
3. Follow the row and column until they meet: the answer is **28**!

📈 My Confidence Tracker 📈

Record your scores below. You'll be amazed at your progress!

My name: _______________________________

✅ Test	📅 Date	⭐ Score	☺ How I Feel
1		/	
2		/	
3		/	

💡 My Quick Reflection

The easiest topic for me was:

____

The trickiest topic for me was:

____

One thing I got better at from Test 1 to Test 3:

____

Next time I want to try:

____

You just finished 3 practice tests — that's awesome! Compare your first score to your last. I bet you'll see real improvement. Ready for more? Check out our 5-test or 7-test books for even more practice!

Get Online

Find more at
ViewMath.com/MN-Grade3

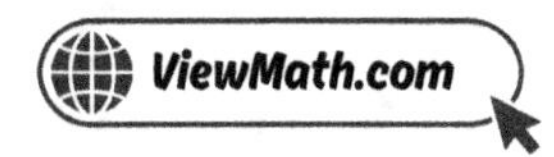

ViewMath.com

Continue Learning at
ViewMath Academy!

For Parents, Teachers & Students

Great job on the practice tests! Want to keep improving? ViewMath Academy is your **free online companion** to this book.

- **Score Analyzer** — Enter your answers and instantly see which topics need more practice

- **Interactive Lessons** — Review the concepts behind each question with clear explanations

- **Adaptive Quizzes** — Practice your weak topics with questions that match your level

- **Progress Tracking** — See your mastery grow across all Grade 3 math topics

- **Personalized Dashboard** — A learning plan tailored just for you

Scan to visit ViewMath Academy

viewmath.com/academy

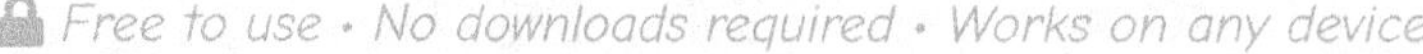 Free to use · No downloads required · Works on any device 

 Table of Contents

Here's what we'll explore together!

 Let's learn and have fun!

Practice Test 1

30 Questions

✏️ Before You Start ✏️

- ✔ **Read each question carefully** before choosing your answer.
- ✔ **Show your work** on scratch paper when you need to.
- ✔ **Skip hard questions** and come back to them later.
- ✔ **Check your answers** when you're done.
- ✔ **Take your time** — there's no rush!

⭐ You've Got This! ⭐

Do your best and show what you know!

1. Write 5,600 in expanded form.

Your Answer:

2. Which number rounds to 500 when rounded to the nearest 100?

   (A) 428

   (B) 449

   (C) 462

   (D) 551

3. Which group contains only even numbers?

   (A) 12, 34, 57

   (B) 20, 46, 88

   (C) 31, 50, 72

   (D) 14, 63, 90

4. A school raised $8,000 last year and $2,000 this year. How much did the school raise in total?

   (A) $6,000

   (B) $10,000

   (C) $82,000

   (D) $28,000

5. What is the value of the digit 8 in 283,041?

   (A) 80

   (B) 800

   (C) 8,000

   (D) 80,000

6. A farmer had 635 apples. He sold 278. How many apples does he have left?

Your Answer

Find more at
ViewMath.com/MN-Grade3

7. *Find the sum:* $8{,}456 + 1{,}544$. *Does the answer have 4 or 5 digits? Explain.*

Your Answer:

8. *Use the standard algorithm to find* $8{,}412 - 3{,}675$.

Your Answer:

9. *If* $6 \times 9 = 54$, *what is* 9×6?

Your Answer:

10. *What is* $56 \div 8$?

(A) 6

(B) 7

(C) 8

(D) 48

11. *In an input/output table, the input is 6 and the output is 30. The rule could be:*

(A) Add 6

(B) Subtract 6

(C) Multiply by 5

(D) Divide by 5

12. *The rule is "Multiply by 8." Which pair of input and output is CORRECT?*

(A) Input $= 5$, Output $= 13$

(B) Input $= 5$, Output $= 40$

(C) Input $= 5$, Output $= 45$

(D) Input $= 5$, Output $= 58$

13. Is $\frac{3}{3}$ at 0 or at 1 on the number line?

Your Answer:

14. What is special about ALL unit fractions?

(A) They all have the same denominator

(B) They all equal 1

(C) They all have 1 as the numerator

(D) They are all less than $\frac{1}{2}$

15. You are at $\frac{3}{6}$ on a number line. You hop 2 more times by $\frac{1}{6}$. Where do you land?

Your Answer:

16. Write 4 as a fraction with denominator 3.

(A) $\frac{4}{3}$

(B) $\frac{3}{4}$

(C) $\frac{7}{3}$

(D) $\frac{12}{3}$

17. Tom ate $\frac{3}{8}$ of a pizza. Mia ate $\frac{5}{8}$ of the same pizza. Who ate more?

(A) Tom

(B) Mia

(C) They ate the same amount

(D) Cannot tell

18. What is $\dfrac{3}{4} + \dfrac{1}{4}$?

Your Answer:

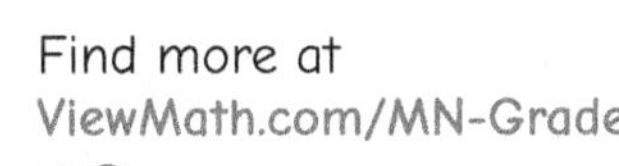
Find more at
ViewMath.com/MN-Grade3

19. When you subtract fractions with the same denominator, what do you do?

(A) Subtract both the numerators and the denom-inators.

(B) Subtract the numerators and keep the denom-inator the same.

(C) Subtract the denominators and keep the nu-merator the same.

(D) Multiply the numerators.

20. A library is open from 10:00 A.M. to 3:00 P.M. How long is it open?

(A) 3 hours

(B) 4 hours

(C) 5 hours

(D) 7 hours

21. An eraser ends at the third quarter-inch mark past 3 inches. How long is the eraser to the nearest quarter inch?

(A) $3\frac{1}{4}$ inches

(B) $3\frac{1}{2}$ inches

(C) $3\frac{3}{4}$ inches

(D) 4 inches

22. A large milk jug holds 4 liters. You pour out 1 liter for cereal. How much milk is left?

Your Answer:

23. How much is a quarter worth?

(A) 1 cent

(B) 5 cents

(C) 10 cents

(D) 25 cents

Find more at
ViewMath.com/MN-Grade3

24. *You buy a book for $4.50 and pay with a $10 bill. How much change do you get?*

(A) $4.50   (B) $5.50

(C) $6.50   (D) $14.50

25. *Look at this picture graph data. Each symbol stands for 1 pet.*

Dogs: ★★★★★★
Cats: ★★★★
Fish: ★★

How many pets are there in all?

(A) 3   (B) 10

(C) 12   (D) 14

26. *A line plot shows the lengths of 10 crayons. The data is shown below:*

2 in.	$2\frac{1}{2}$ in.	3 in.	$3\frac{1}{2}$ in.	4 in.
2	3	4	1	0

How many crayons are shorter than 3 inches?

(A) 2   (B) 4

(C) 5   (D) 9

27. *Is it possible to roll a number greater than 6 on a regular die numbered 1 through 6?*

Your Answer:

Find more at
ViewMath.com/MN-Grade3

28. *What unit do we use to measure area?*

    (A) Centimeters

    (B) Square units

    (C) Inches

    (D) Pounds

29. *Lily says a rectangle that is 5 cm long and 3 cm wide has a perimeter of 15 cm. What mistake did she make?*

    (A) She subtracted instead of adding.

    (B) She found the area instead of the perimeter.

    (C) She forgot to add all 4 sides.

    (D) She divided instead of multiplying.

30. *If one pizza is cut into 4 equal slices and another identical pizza is cut into 8 equal slices, which slices are bigger?*

    (A) The $\frac{1}{8}$ slices

    (B) The $\frac{1}{4}$ slices

    (C) They are the same size.

    (D) You cannot tell.

Find more at
ViewMath.com/MN-Grade3

 # End of Practice Test 1 

Great job finishing the test!

 My Score

I got _____________ out of 30 questions right.

Check your answers in the **Answer Key** at the back of the book.

Review any questions you missed. That's how we learn!

Check Your Score Online!

Visit **ViewMath Academy** to enter your answers and see which topics you need to review. You can also explore lessons, take quizzes, track your scores, and save your progress!

viewmath.com/score/3.1.MN.01

Or go to viewmath.com/score and enter code: 3.1.MN.01

2

Practice Test 2

 30 Questions

✏ Before You Start ✏

- ✔ **Read each question carefully** before choosing your answer.
- ✔ **Show your work** on scratch paper when you need to.
- ✔ **Skip hard questions** and come back to them later.
- ✔ **Check your answers** when you're done.
- ✔ **Take your time** — there's no rush!

⭐ You've Got This! ⭐

Do your best and show what you know!

1. *Write the expanded form of 9,205.*

Your Answer:

2. *Maria rounded 438 to the nearest 10 and got 430. Is she correct?*

(A) No, it should be 440

(B) Yes, 438 rounds to 430

(C) No, it should be 400

(D) No, it should be 450

3. *List all the even numbers between 21 and 30.*

Your Answer:

4. *How many hundreds are in 10,000?*

Your Answer:

5. *Which is the correct expanded form of 720,056?*

(A) $72,000 + 56$

(B) $700,000 + 20,000 + 50 + 6$

(C) $700,000 + 2,000 + 56$

(D) $700,000 + 20,000 + 56$

6. *What is $413 - 268$?*

(A) 155

(B) 245

(C) 145

(D) 235

Find more at
ViewMath.com/MN-Grade3

7. Can two 4-digit numbers add up to a 5-digit number?

   (A) No, never.

   (B) Yes, when the thousands column has a carry.

   (C) Only with regrouping in every column.

   (D) Only if both numbers are 9,999.

8. The standard algorithm for subtraction works for numbers of any size because it is based on:

   (A) Memorizing answers

   (B) Place value and regrouping

   (C) Estimation

   (D) Multiplication

9. If $7 \times 9 = 63$, what is 9×7?

   (A) 16

   (B) 56

   (C) 63

   (D) 79

10. In the division sentence $48 \div 6 = 8$, what is the divisor?

    (A) 48

    (B) 6

    (C) 8

    (D) 54

11. What is the next number in this pattern? $9, 18, 27, 36, \ldots$

    (A) 40

    (B) 42

    (C) 44

    (D) 45

12. Look at the table. What is the rule?

Input	Output
4	16
7	28
10	40

(A) Add 12

(B) Multiply by 3

(C) Multiply by 4

(D) Add 14

13. On a number line divided into 8 parts, what fraction names the fifth mark from 0?

(A) $\frac{5}{8}$

(B) $\frac{8}{5}$

(C) $\frac{3}{8}$

(D) $\frac{5}{5}$

14. Which list shows unit fractions from smallest to largest?

(A) $\frac{1}{2}, \frac{1}{4}, \frac{1}{8}$

(B) $\frac{1}{8}, \frac{1}{4}, \frac{1}{2}$

(C) $\frac{1}{4}, \frac{1}{2}, \frac{1}{8}$

(D) $\frac{1}{8}, \frac{1}{2}, \frac{1}{4}$

15. Alex says $\frac{3}{4}$ is 3 copies of $\frac{1}{3}$. Is Alex correct?

(A) Yes

(B) No, $\frac{3}{4}$ is 3 copies of $\frac{1}{4}$

(C) No, $\frac{3}{4}$ is 4 copies of $\frac{1}{3}$

(D) No, $\frac{3}{4}$ is 3 copies of $\frac{3}{3}$

16. Which of these is a way to write 5 as a fraction?

(A) $\frac{5}{1}$

(B) $\frac{10}{2}$

(C) $\frac{15}{3}$

(D) All of the above

Find more at
ViewMath.com/MN-Grade3

17. Is $\frac{5}{8}$ more or less than $\frac{1}{2}$?

    (A) Less than $\frac{1}{2}$ (B) Equal to $\frac{1}{2}$

    (C) More than $\frac{1}{2}$ (D) Cannot tell

18. What is $\frac{1}{4} + \frac{1}{4} + \frac{1}{4}$?

    Your Answer:

19. What is $\frac{7}{8} - \frac{4}{8}$?

    Your Answer:

20. A TV show starts at 4:15 P.M. and ends at 5:00 P.M. How long is the TV show?

    Your Answer:

21. Which measurement is the same as $\frac{1}{2}$ inch?

    (A) $\frac{1}{4}$ inch   (B) $\frac{2}{4}$ inch

    (C) $\frac{3}{4}$ inch   (D) 1 inch

22. Container A holds 20 liters. Container B holds 13 liters. How much more does Container A hold?

    (A) 7 L   (B) 13 L

    (C) 20 L   (D) 33 L

Find more at
ViewMath.com/MN-Grade3

ViewMath.com

23. *Liam buys an apple for $0.85 and a banana for $0.40. He pays with a $5 bill. How much change does he get?*

(A) $3.25

(B) $3.75

(C) $4.15

(D) $4.60

24. *Subtract: $12.00 − $7.45*

Your Answer:

25. *Lily says the picture graph below shows that 4 kids were on the swings.*

Swings: ★★★★
Slide: ★★★★★★
Key: Each ★ = 2 *kids*

What is the correct number of kids on the swings?

(A) 2

(B) 4

(C) 6

(D) 8

26. *Leo measured 8 worms. His line plot shows 7 X marks total. What went wrong?*

(A) *He measured one worm twice*

(B) *He forgot to plot one measurement*

(C) *He used the wrong numbers on the number line*

(D) *Nothing — 7 is correct*

27. *Is the following event certain, likely, unlikely, or impossible?*
"Tomorrow will come after today."

Your Answer:

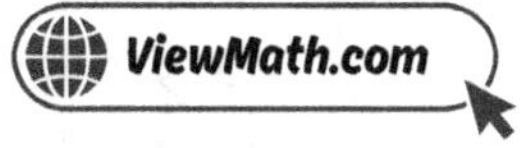

28. A shape is covered with unit squares. You count 3 rows with 8 squares in each row. What is the area?

Your Answer:

29. A triangle has sides of 4 cm, 6 cm, and 8 cm. What is the perimeter?

(A) 14 cm

(B) 18 cm

(C) 20 cm

(D) 24 cm

30. You fold a piece of paper in half, then fold it in half again. How many equal parts do you have?

Your Answer:

Find more at
ViewMath.com/MN-Grade3

Great job finishing the test!

✅ My Score

I got _____________ out of 30 questions right.

*Check your answers in the **Answer Key** at the back of the book.*

💡 *Review any questions you missed. That's how we learn!*

📊 Check Your Score Online!

Visit **ViewMath Academy** to enter your answers and see which topics you need to review. You can also explore lessons, take quizzes, track your scores, and save your progress!

viewmath.com/score/3.1.MN.02

Or go to viewmath.com/score and enter code: 3.1.MN.02

Practice Test 3

 30 Questions

✏ Before You Start ✏

- ✔ **Read each question carefully** before choosing your answer.
- ✔ **Show your work** on scratch paper when you need to.
- ✔ **Skip hard questions** and come back to them later.
- ✔ **Check your answers** when you're done.
- ✔ **Take your time** — there's no rush!

 You've Got This! 

Do your best and show what you know!

1. Each place value is how many times bigger than the place to its right?

   (A) 2 times   (B) 5 times

   (C) 10 times   (D) 100 times

2. When you round 549 to the nearest 10 and to the nearest 100, which gives the larger answer?

   (A) Rounded to the nearest 10   (B) Rounded to the nearest 100

   (C) They are the same   (D) Cannot tell without rounding

3. If you add two odd numbers, the result is always:

   (A) Odd   (B) Even

   (C) Greater than 10   (D) An odd number less than 20

4. What number is 1 more than 9,999?

   (A) 9,000   (B) 9,990

   (C) 10,000   (D) 10,001

5. What is the smallest 6-digit number?

   Your Answer:

6. What is 524 − 367 using the standard algorithm?

   (A) 157   (B) 167

   (C) 257   (D) 147

Find more at
ViewMath.com/MN-Grade3

7. *Maya added $4,839 + 2,764$ and got $7,503$. She estimates $4,800 + 2,800 = 7,600$. What should Maya do?*

(A) *Accept her answer since it is close to the esti-mate.*

(B) *Redo the problem because the estimate is too far away.*

(C) *Round both numbers to the nearest thousand instead.*

(D) *Subtract to check.*

8. *What is the trickiest part of the standard algorithm when subtracting from a number like $4,000$?*

(A) *There are too many digits.*

(B) *You cannot subtract from round numbers.*

(C) *You must borrow across multiple zeros.*

(D) *You need to add instead.*

9. *Which property is used here?* $8 \times 7 = 8 \times (5 + 2) = (8 \times 5) + (8 \times 2)$

(A) *Commutative property*

(B) *Associative property*

(C) *Distributive property*

(D) *Zero property*

10. *Carlos starts with 40 stickers. He gives away 5 stickers at a time. How many times can he give away stickers?*

(A) 7

(B) 8

(C) 9

(D) 35

11. *Write the next three numbers in this pattern and state the rule:* $6, 12, 18, \ldots$

Your Answer:

12. Look at the table. What is the rule?

Input	Output
10	4
12	6
15	9

Your Answer:

13. A number line from 0 to 1 is divided into 6 equal parts. Which fraction is closest to 1?

(A) $\frac{1}{6}$

(B) $\frac{3}{6}$

(C) $\frac{4}{6}$

(D) $\frac{5}{6}$

14. Which list shows unit fractions in order from largest to smallest?

(A) $\frac{1}{8}, \frac{1}{6}, \frac{1}{4}, \frac{1}{3}$

(B) $\frac{1}{3}, \frac{1}{4}, \frac{1}{6}, \frac{1}{8}$

(C) $\frac{1}{4}, \frac{1}{8}, \frac{1}{3}, \frac{1}{6}$

(D) $\frac{1}{2}, \frac{1}{8}, \frac{1}{4}, \frac{1}{6}$

15. Count by fourths: $\frac{1}{4}, \frac{2}{4}, \underline{\quad}, \frac{4}{4}$. What fraction is missing?

Your Answer:

16. On a number line divided into fourths, which fraction is at 3?

(A) $\frac{3}{4}$

(B) $\frac{4}{3}$

(C) $\frac{12}{4}$

(D) $\frac{3}{1}$

17. When two fractions have the same denominator, how do you compare them?

   (A) The one with the bigger denominator is larger   (B) The one with the bigger numerator is larger

   (C) They are always equal   (D) You cannot compare them

18. What is $\frac{2}{3} + \frac{1}{3}$?

   (A) $\frac{3}{6}$ (B) $\frac{3}{3}$

   (C) $\frac{2}{3}$ (D) $\frac{1}{3}$

19. Liam says $\frac{8}{10} - \frac{3}{10} = \frac{5}{0}$. What did Liam do wrong?

   (A) He added instead of subtracting.   (B) He subtracted the denominators too.

   (C) He used the wrong numerators.   (D) He did nothing wrong.

20. Emma started reading at 3:30 and stopped at 4:10. How long did she read?

   (A) 30 minutes   (B) 40 minutes

   (C) 50 minutes   (D) 1 hour and 10 minutes

21. A pencil is $7\frac{1}{4}$ inches long. A crayon is $5\frac{3}{4}$ inches long. How much longer is the pencil than the crayon?

   Your Answer:

22. What unit do we use to measure liquid volume?

   (A) Grams   (B) Kilograms

   (C) Liters   (D) Inches

Find more at
ViewMath.com/MN-Grade3

ViewMath.com

23. How many cents are in $1.00?

(A) 1 cent   (B) 10 cents

(C) 50 cents   (D) 100 cents

24. What is $4.60 + $3.85?

(A) $7.45   (B) $8.45

(C) $8.35   (D) $7.85

25. A picture graph tracks books read in a month. Each symbol stands for 2 books. Amy has 5 symbols, Ben has 8 symbols, and Chloe has 3 symbols. How many books did they read altogether?

Your Answer:

26. Which type of graph is BEST for showing measurement data like the lengths of worms in inches?

(A) Picture graph   (B) Bar graph

(C) Line plot   (D) Tally chart

27. Which event is certain?

(A) You will roll a 3 on a die.   (B) The day after Monday is Tuesday.

(C) It will snow tomorrow.   (D) You will pick a red card from a deck.

28. What is a unit square?

(A) A square with sides that are 10 units long   (B) A square with sides that are 1 unit long

(C) Any square shape   (D) A rectangle with 4 equal sides

29. *Which shape has the greatest perimeter?*

(A) A square with sides of 5 cm

(B) A rectangle that is 8 cm × 3 cm

(C) A rectangle that is 6 cm × 4 cm

(D) A triangle with sides 7 cm, 7 cm, and 7 cm

30. *A square is partitioned into 4 equal parts. All 4 parts are shaded. What fraction is shaded?*

(A) $\frac{1}{4}$

(B) $\frac{3}{4}$

(C) $\frac{4}{4}$

(D) $\frac{4}{1}$

Find more at
ViewMath.com/MN-Grade3

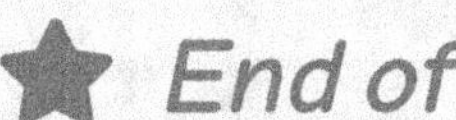 

⭐ End of Practice Test 3 ⭐

Great job finishing the test!

📋 My Score

I got ______________ out of 30 questions right.

*Check your answers in the **Answer Key** at the back of the book.*

💡 *Review any questions you missed. That's how we learn!*

📊 Check Your Score Online!

Visit **ViewMath Academy** to enter your answers and see which topics you need to review. You can also explore lessons, take quizzes, track your scores, and save your progress!

viewmath.com/score/3.1.MN.03

Or go to *viewmath.com/score* and enter code: **3.1.MN.03**

Answer Key & Explanations

★ Check Your Answers! ★

First try each test on your own, then look here to check.

Read the explanations to learn from any mistakes ★

☑ Practice Test 1 — Answer Key

1 $5,000 + 600$ **2** C   **3** B   **4** B   **5** D   **6** 357

7 10,000. *It has 5 digits because the thousands column produces a carry.*   **8** 4,737   **9** 54   **10** B

11 C   **12** B   **13** 1   **14** C   **15** $\frac{5}{6}$ **16** D   **17** B   **18** $\frac{4}{4}$ or 1   **19** B   **20** C

21 C   **22** 3 L   **23** D   **24** B   **25** C   **26** C   **27** *No, it is impossible*   **28** B

29 B   **30** B

💡 Time to Learn! 💡

Go through the explanations below, **especially for the questions you missed**.

Understanding why each answer is correct makes you a stronger math thinker!

👍 **Tip:** Circle any questions you got wrong, then read their explanation carefully.

📖 Practice Test 1 — Detailed Explanations

Find more at
ViewMath.com/MN-Grade3

1. $5{,}600 = 5{,}000 + 600 + 0 + 0$. The tens and ones are both 0.

2. 462 has tens digit $6 \geq 5$, so it rounds up to 500. 428 and 449 round to 400. 551 rounds to 600.

3. 20 (ends in 0), 46 (ends in 6), 88 (ends in 8) are all even. The other groups each contain at least one odd number.

4. $8{,}000 + 2{,}000 = 10{,}000$.

5. In 283,041, the digit 8 is in the ten-thousands place, so its value is 80,000.

6. Ones: $5 < 8$. Borrow: $15 - 8 = 7$. Tens: $2 < 7$. Borrow: $12 - 7 = 5$. Hundreds: $5 - 2 = 3$. The farmer has 357 apples left.

7. Ones: $6 + 4 = 10$, write 0, carry 1. Tens: $5 + 4 + 1 = 10$, write 0, carry 1. Hundreds: $4 + 5 + 1 = 10$, write 0, carry 1. Thousands: $8 + 1 + 1 = 10$. Sum: 10,000, a 5-digit number.

8. Ones: $2 < 5$, borrow. $12 - 5 = 7$. Tens: $0 < 7$, borrow. $10 - 7 = 3$. Hundreds: $3 - 6$? Borrow. $13 - 6 = 7$. Thousands: $7 - 3 = 4$. Difference: 4,737.

9. By the commutative property, swapping the factors does not change the product.

10. $56 \div 8 = 7$ because $8 \times 7 = 56$.

11. $6 \times 5 = 30$. The rule is "multiply by 5." You should check this with other rows to be sure.

12. $5 \times 8 = 40$. The correct output for input 5 is 40.

13. $\frac{3}{3}$ means all 3 parts, which equals 1 whole.

14 A unit fraction always has 1 as the numerator. Examples: $\frac{1}{2}, \frac{1}{3}, \frac{1}{4}, \frac{1}{8}$.

15 $\frac{3}{6} + \frac{1}{6} + \frac{1}{6} = \frac{5}{6}$.

16 $4 \times 3 = 12$ thirds. So $4 = \frac{12}{3}$.

17 Same denominator. $5 > 3$, so $\frac{5}{8} > \frac{3}{8}$. Mia ate more.

18 $\dfrac{3+1}{4} = \dfrac{4}{4} = 1$ whole.

19 To subtract fractions with like denominators, subtract the numerators and keep the denominator the same.

20 From 10:00 A.M. to 3:00 P.M.: count $10 \to 11 \to 12 \to 1 \to 2 \to 3$, which is 5 hours.

21 The third quarter-inch mark past 3 is $3\frac{3}{4}$ inches.

22 $4 - 1 = 3$ L.

23 A quarter is worth 25 cents.

24 $\$10.00 - \$4.50 = \$5.50$.

25 Dogs: 6, Cats: 4, Fish: 2. Total: $6 + 4 + 2 = 12$ pets.

26 Crayons shorter than 3 inches: 2 in. (2) + $2\frac{1}{2}$ in. (3) = 5 crayons.

27 A regular die only has numbers 1 through 6, so rolling a number greater than 6 is impossible.

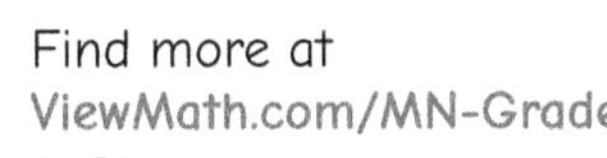
Find more at
ViewMath.com/MN-Grade3

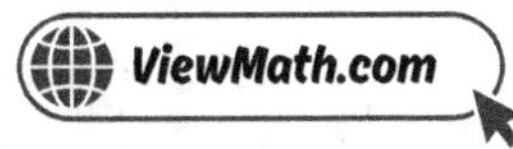

28) Area is measured in square units (such as square centimeters, square inches, etc.), not regular length units.

29) Lily multiplied $5 \times 3 = 15$, which gives the area, not the perimeter. The correct perimeter is $5 + 3 + 5 + 3 = 16$ cm.

30) The $\frac{1}{4}$ slices are bigger because the pizza is cut into fewer pieces. More equal parts means smaller pieces: $\frac{1}{4} > \frac{1}{8}$.

☑ Practice Test 2 — Answer Key

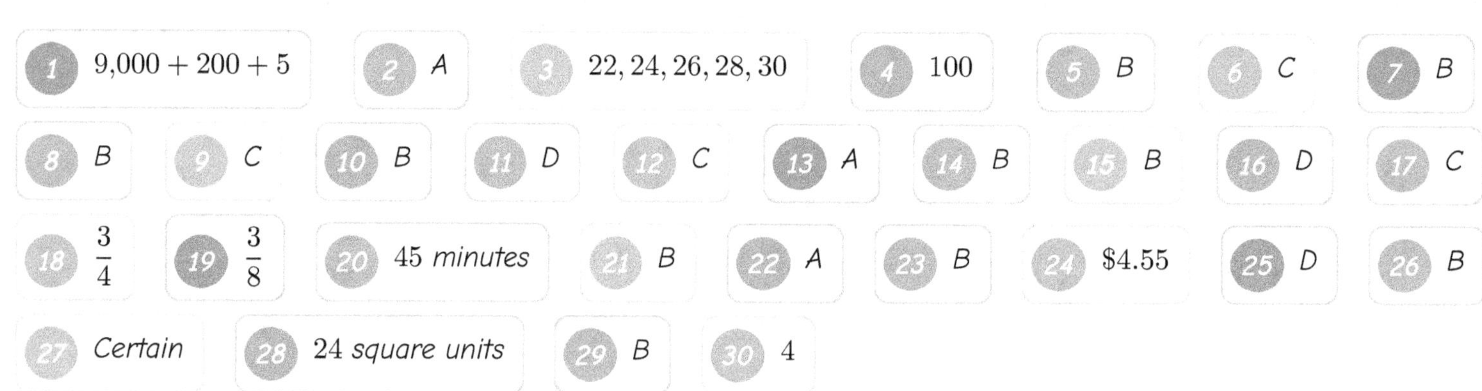

1) $9,000 + 200 + 5$ 2) A   3) $22, 24, 26, 28, 30$ 4) 100   5) B   6) C   7) B

8) B   9) C   10) B   11) D   12) C   13) A   14) B   15) B   16) D   17) C

18) $\frac{3}{4}$ 19) $\frac{3}{8}$ 20) 45 minutes   21) B   22) A   23) B   24) $4.55   25) D   26) B

27) Certain   28) 24 square units   29) B   30) 4

💡 Time to Learn! 💡

Go through the explanations below, **especially for the questions you missed.**

Understanding why each answer is correct makes you a stronger math thinker!

👍 **Tip:** Circle any questions you got wrong, then read their explanation carefully.

📖 Practice Test 2 — Detailed Explanations

1) $9,205 = 9,000 + 200 + 0 + 5$. The tens digit is 0, so there is no tens term.

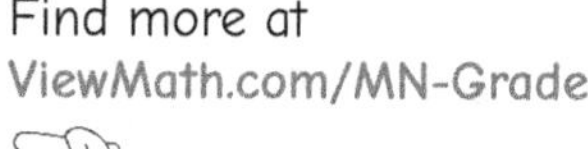

2 The ones digit is 8. Since $8 \geq 5$, we round up. 438 rounds to 440, not 430.

3 The even numbers between 21 and 30 are $22, 24, 26, 28, 30$. They all end in an even digit.

4 $10{,}000 \div 100 = 100$. There are 100 hundreds in 10,000.

5 $720{,}056 = 700{,}000 + 20{,}000 + 0 + 0 + 50 + 6$. The thousands and hundreds places are both zero.

6 Ones: $3 < 8$. Borrow: $13 - 8 = 5$. Tens: $0 < 6$. Borrow: $10 - 6 = 4$. Hundreds: $3 - 2 = 1$. The difference is 145.

7 Yes! If the thousands column (plus any carry) totals 10 or more, the sum is a 5-digit number. For example, $6{,}500 + 4{,}500 = 11{,}000$.

8 The standard algorithm works for any number of digits because it is based on **place value** — subtracting column by column — and **regrouping** (borrowing) when needed.

9 By the commutative property, $9 \times 7 = 7 \times 9 = 63$.

10 The divisor is the number you divide by. In $48 \div 6 = 8$, the divisor is 6.

11 The pattern adds 9 each time: $36 + 9 = 45$. Digit check: $4 + 5 = 9$. ✓

12 $4 \times 4 = 16$, $7 \times 4 = 28$, $10 \times 4 = 40$. The rule is multiply by 4.

13 Counting 5 marks from 0 on a line divided into 8 parts gives $\frac{5}{8}$.

14 Bigger denominators mean smaller pieces: $\frac{1}{8} < \frac{1}{4} < \frac{1}{2}$.

Find more at
ViewMath.com/MN-Grade3

15 $\frac{3}{4}$ means 3 copies of $\frac{1}{4}$, not $\frac{1}{3}$. The denominator tells the size of each piece.

16 $\frac{5}{1} = 5$, $\frac{10}{2} = 5$, and $\frac{15}{3} = 5$. They all equal 5.

17 $\frac{1}{2} = \frac{4}{8}$. Since $\frac{5}{8} > \frac{4}{8}$, it is more than $\frac{1}{2}$.

18 $\dfrac{1+1+1}{4} = \dfrac{3}{4}$.

19 $\dfrac{7-4}{8} = \dfrac{3}{8}$.

20 From 4:15 to 5:00: count 45 minutes forward. Or $60 - 15 = 45$ minutes.

21 $\frac{2}{4} = \frac{1}{2}$, so $\frac{2}{4}$ inch is the same as $\frac{1}{2}$ inch.

22 $20 - 13 = 7$ L.

23 Total: $\$0.85 + \$0.40 = \$1.25$. Change: $\$5.00 - \$1.25 = \$3.75$.

24 $\$12.00 - \$7.45 = \$4.55$. Borrow to subtract 45 cents from 0 cents.

25 Lily forgot to multiply by the key. Each star = 2 kids, and there are 4 stars: $4 \times 2 = 8$ kids.

26 He measured 8 worms but only has 7 X marks, so he forgot to plot one measurement.

27 Tomorrow always comes after today. This will definitely happen.

28 3 rows $\times$ 8 columns = 24 square units.

29 Add all side lengths: $4 + 6 + 8 = 18$ cm.

30 Folding in half gives 2 parts. Folding in half again doubles it to 4 equal parts. Each part is $\frac{1}{4}$ of the whole.

✅ Practice Test 3 — Answer Key

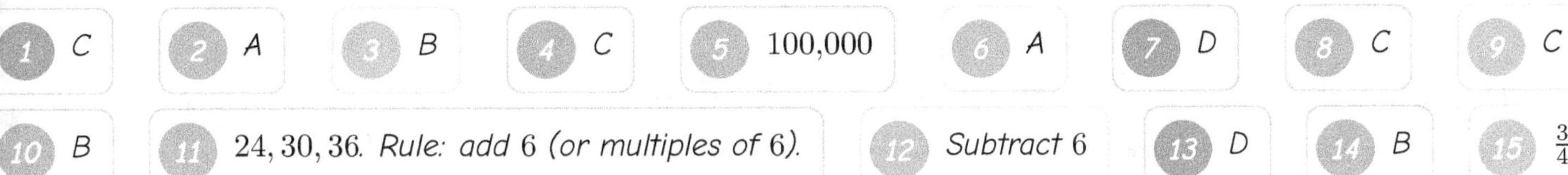

| 1 C | 2 A | 3 B | 4 C | 5 100,000 | 6 A | 7 D | 8 C | 9 C |

| 10 B | 11 24, 30, 36. Rule: add 6 (or multiples of 6). | 12 Subtract 6 | 13 D | 14 B | 15 $\frac{3}{4}$ |

| 16 C | 17 B | 18 B | 19 B | 20 B | 21 $1\frac{1}{2}$ inches | 22 C | 23 D | 24 B |

| 25 32 | 26 C | 27 B | 28 B | 29 B | 30 C |

💡 Time to Learn! 💡

*Go through the explanations below, **especially for the questions you missed**.*

Understanding why each answer is correct makes you a stronger math thinker!

👍 **Tip:** Circle any questions you got wrong, then read their explanation carefully.

📖 Practice Test 3 — Detailed Explanations

1 Each place is 10 *times* bigger than the place to its right: ones → tens → hundreds → thousands.

2 549 rounded to the nearest 10: ones digit is $9 \geq 5$, so 550. Rounded to the nearest 100: tens digit is $4 < 5$, so 500. $550 > 500$.

3) Odd + Odd = Even. For example, $3 + 5 = 8$ (even) and $7 + 9 = 16$ (even).

4) $9,999 + 1 = 10,000$. This is the first 5-digit number.

5) The smallest 6-digit number is $100,000$. Any number below it (like $99,999$) has only 5 digits.

6) Ones: $4 < 7$. Borrow: $14 - 7 = 7$. Tens: $1 < 6$. Borrow: $11 - 6 = 5$. Hundreds: $4 - 3 = 1$. The difference is 157.

7) The estimate $7,600$ is close to $7,503$. To be sure, Maya should check by subtracting: $7,503 - 2,764$ should equal $4,839$. The correct sum is actually $7,603$, so she has an error.

8) When there are zeros in the tens and hundreds places, you cannot borrow directly. You must keep going left to find a non-zero digit and pass 10s back through each zero column.

9) Breaking apart a factor and multiplying each part is the distributive property.

10) $40 \div 5 = 8$. He can give away stickers 8 times. This is like subtracting 5 from 40 repeatedly until reaching 0, which takes 8 steps.

11) Each number increases by 6: $18 + 6 = 24$, $24 + 6 = 30$, $30 + 6 = 36$.

12) $10 - 6 = 4$, $12 - 6 = 6$, $15 - 6 = 9$. The rule is subtract 6.

13) $\frac{5}{6}$ is the last mark before 1, so it is closest to 1.

14) Largest to smallest: bigger pieces first. $\frac{1}{3} > \frac{1}{4} > \frac{1}{6} > \frac{1}{8}$.

15) Counting by $\frac{1}{4}$: $\frac{1}{4}, \frac{2}{4}, \frac{3}{4}, \frac{4}{4}$.

Find more at
ViewMath.com/MN-Grade3

16 $3 = 3 \times \frac{4}{4} = \frac{12}{4}$. On a fourths number line, 3 is at $\frac{12}{4}$.

17   Same denominator means same-size pieces. More pieces (bigger numerator) = bigger fraction.

18 $\frac{2+1}{3} = \frac{3}{3}$, which equals 1 whole.

19   Liam subtracted both the numerators and the denominators. The denominator should stay the same: $\frac{8-3}{10} = \frac{5}{10}$.

20   From 3:30 to 4:00 is 30 minutes. From 4:00 to 4:10 is 10 minutes. Total: $30 + 10 = 40$ minutes.

21 $7\frac{1}{4} - 5\frac{3}{4} = 6\frac{5}{4} - 5\frac{3}{4} = 1\frac{2}{4} = 1\frac{1}{2}$ inches.

22   We measure liquid volume in liters (L).

23 $\$1.00 = 100$ cents.

24   Cents: $60 + 85 = 145$ cents $= 1$ dollar and 45 cents. Dollars: $4 + 3 + 1 = 8$. Answer: $\$8.45$.

25   Amy: $5 \times 2 = 10$. Ben: $8 \times 2 = 16$. Chloe: $3 \times 2 = 6$. Total: $10 + 16 + 6 = 32$ books.

26   Line plots are best for measurement data. They show each individual measurement as an X on a number line, making it easy to see how values are spread out.

27   Tuesday always comes after Monday. This will definitely happen, so it is certain.

28   A unit square is a square with sides that are 1 unit long. We use unit squares to measure area.

Find more at
ViewMath.com/MN-Grade3

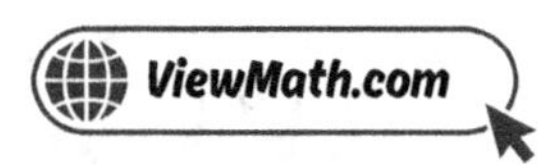

29. Square: $4 \times 5 = 20$ cm. Rectangle 8×3: $(2 \times 8) + (2 \times 3) = 22$ cm. Rectangle 6×4: $(2 \times 6) + (2 \times 4) = 20$ cm. Triangle: $7 + 7 + 7 = 21$ cm. The 8×3 rectangle has the greatest perimeter of 22 cm.

30. All 4 out of 4 parts are shaded, so $\frac{4}{4}$ is shaded. $\frac{4}{4} = 1$ whole.

Great job checking your work!

Keep practicing and you'll be a math star!

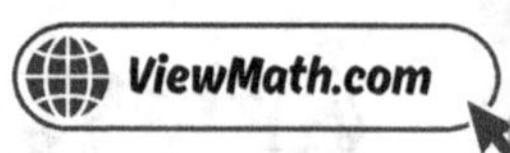